For Mama, Dad, and Lily

Have you ever wondered how your body gets the nutrients it needs?

There are little helpers that support you with everyday life!

Imagine a huge population of tiny creatures living inside and on your body!

This community of organisms is called your microbiome. But who are they?

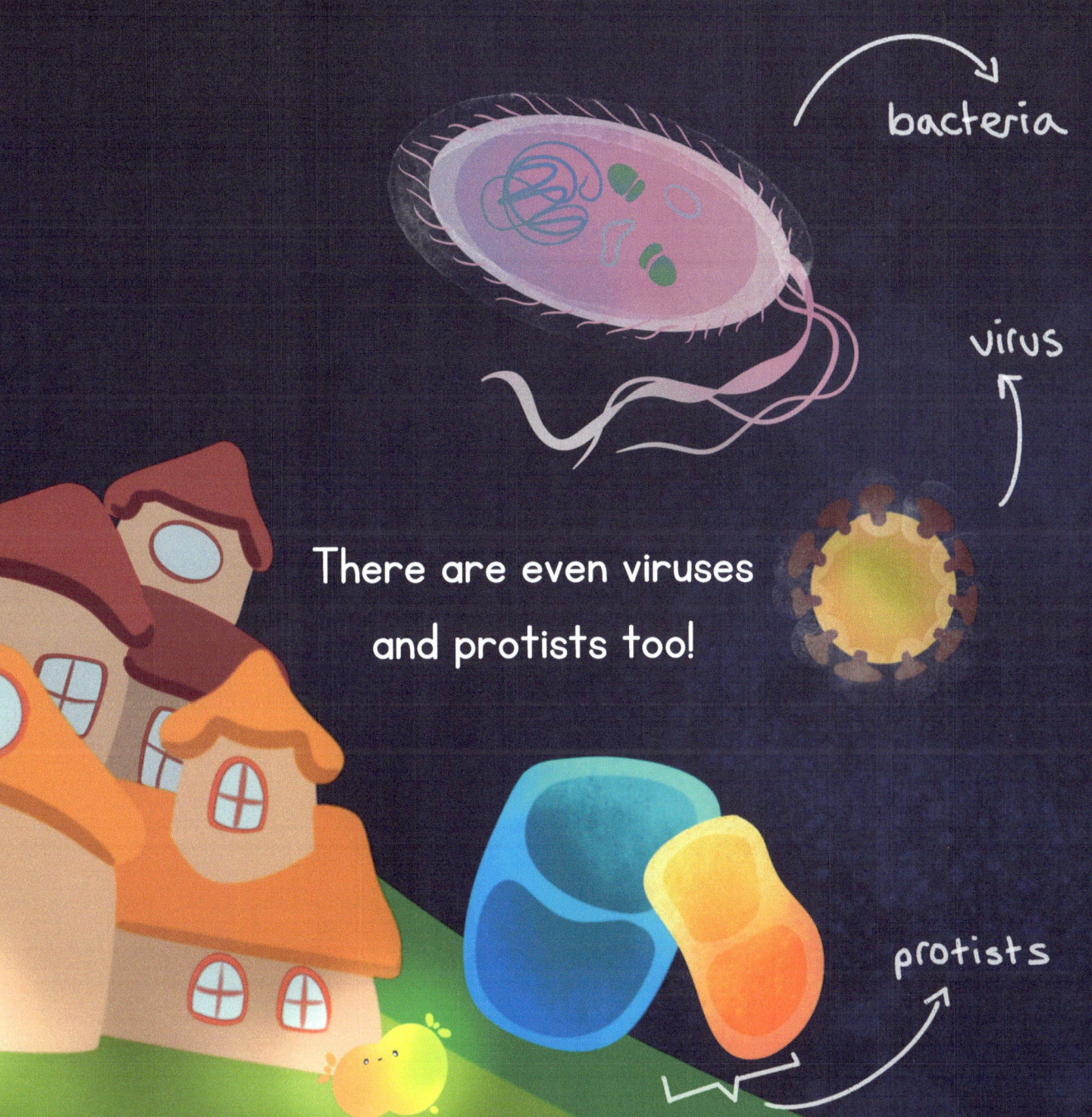

Your microbiome starts to form even before you are born! When you are a baby, you get your first microbes from your mom.

In exchange for giving these creatures a safe place to live and food to eat, they work for us.

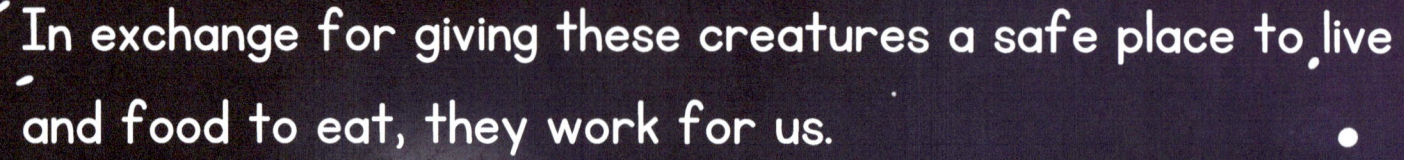

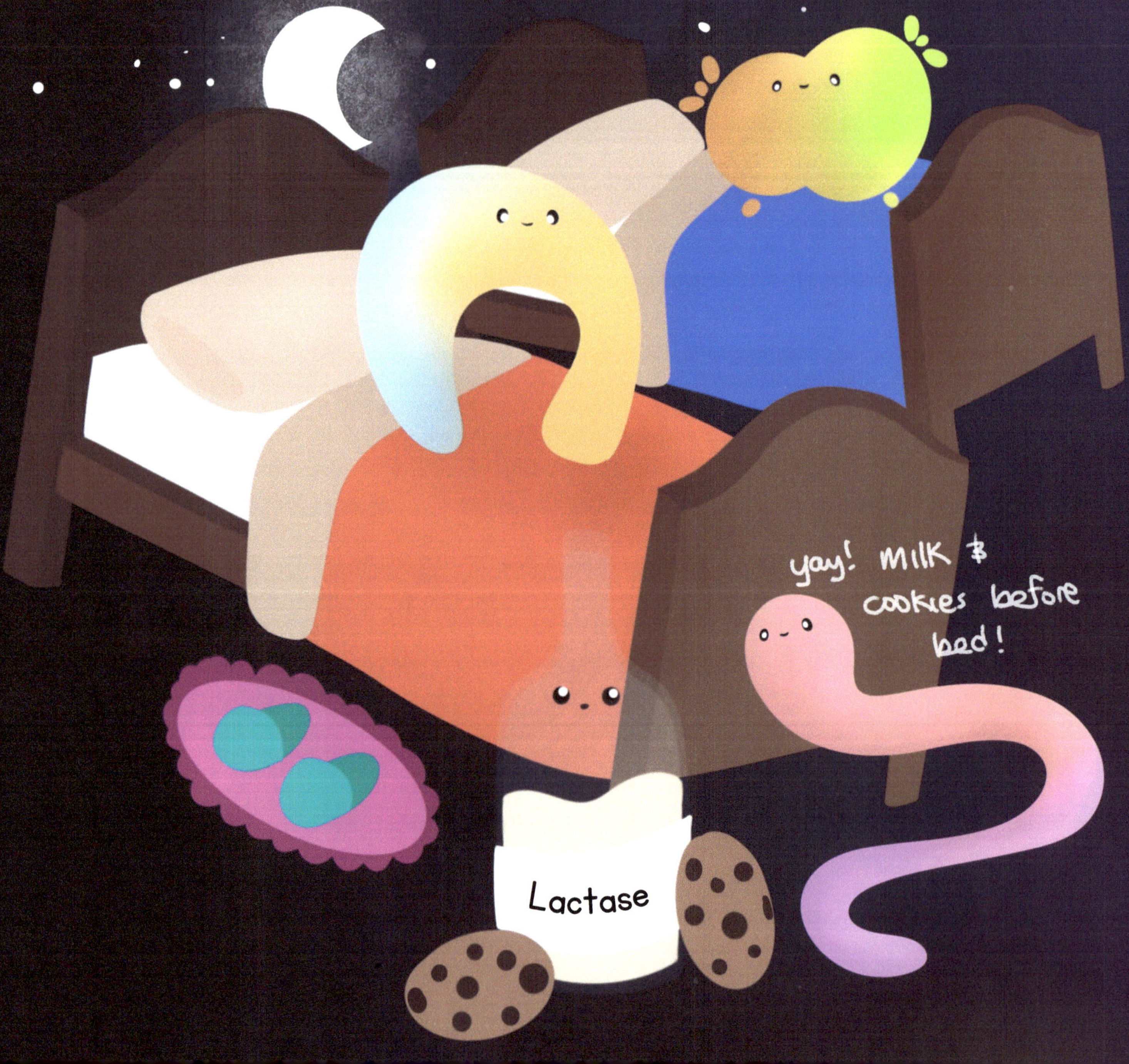

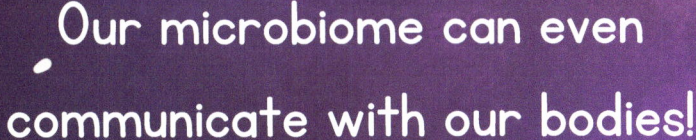

And, since the brain decides what we eat, the microbiome wants a healthy brain. Your microbiome can even motivate you to exercise!

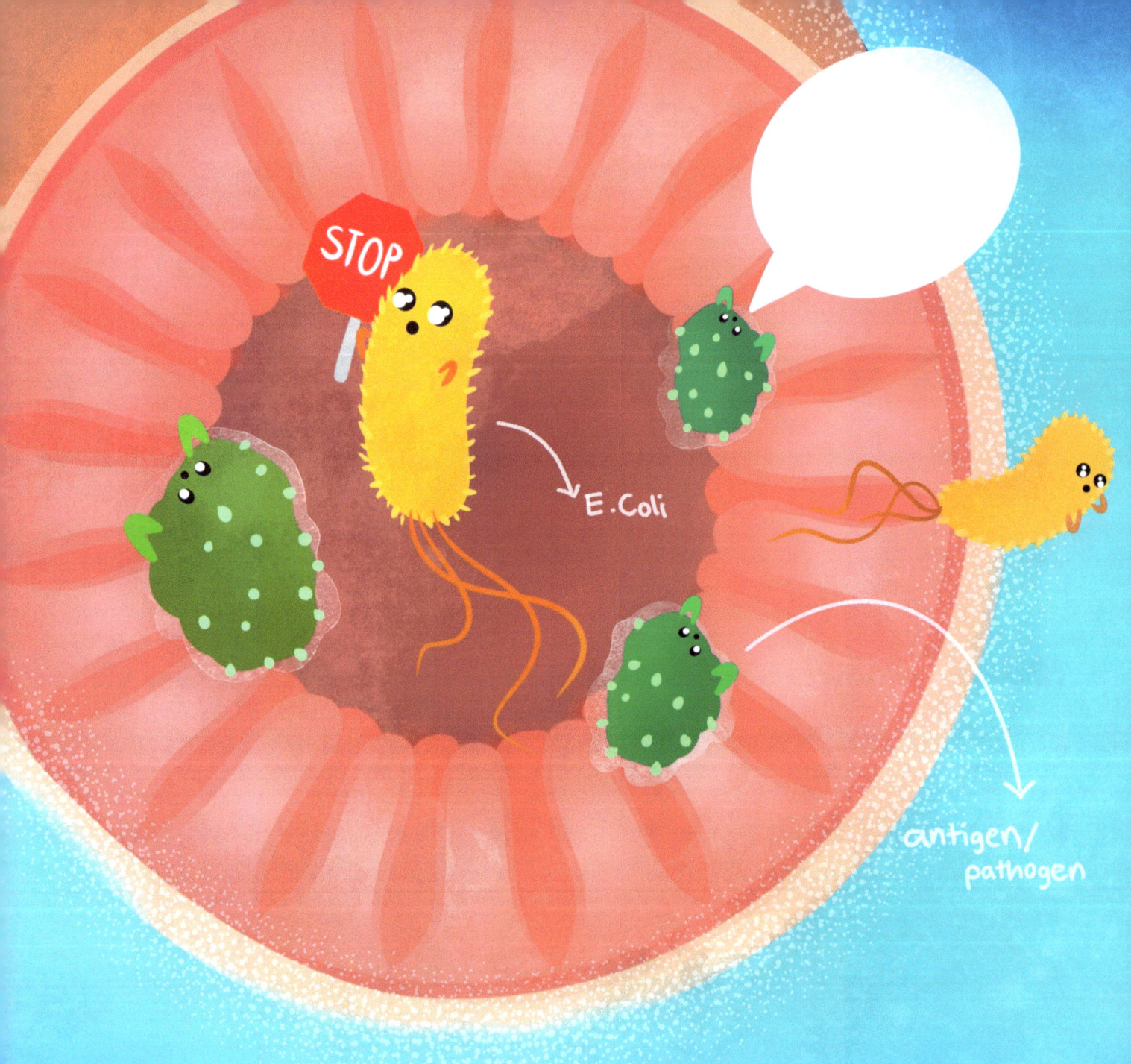

These bacteria also prevent, bad, disease-causing bacteria from sticking to our intestinal walls.

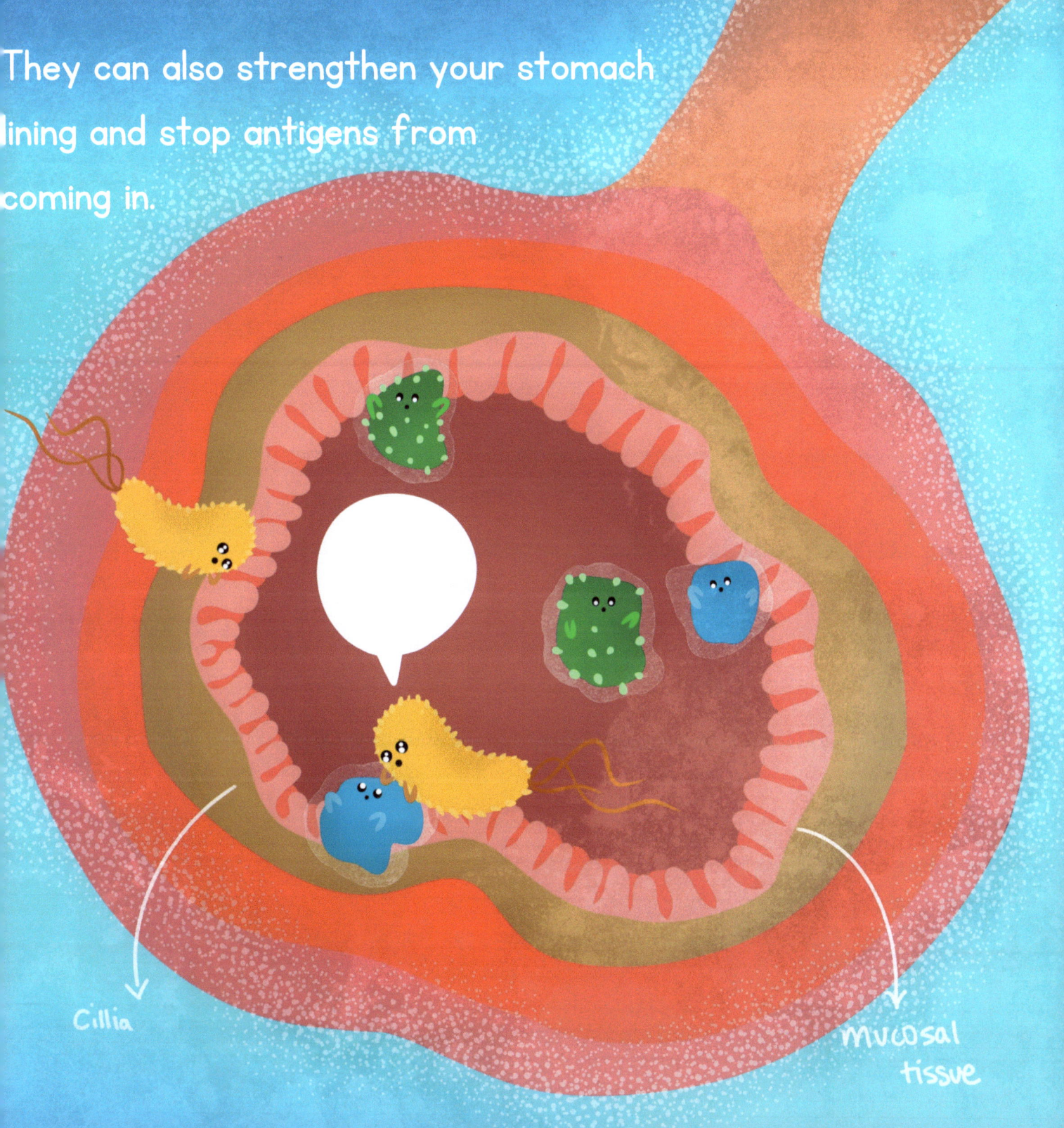

Some like fibers and leafy greens, others go for sugars and starches, and some love oily bacon and fried chicken.

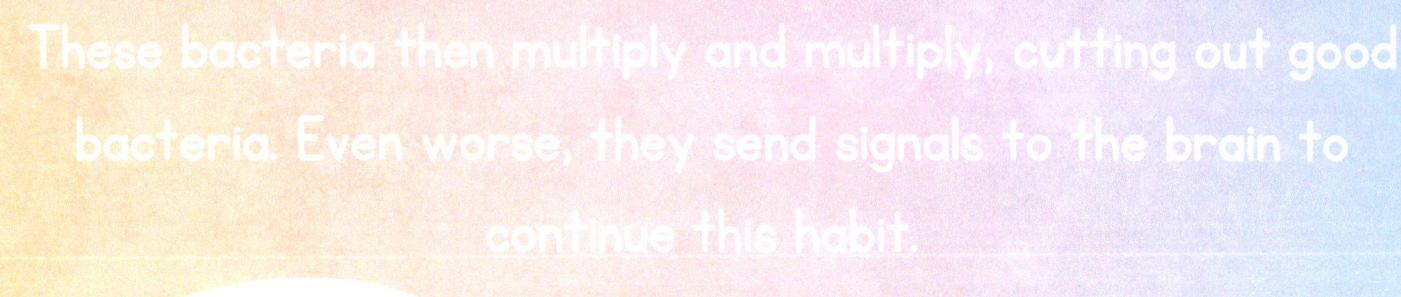

Your microbiome can also affect your mood!

Its important to use antibiotics only when you need to, and eat food with probiotics like yogurt to restore the good microbes in your body!

You can be a great teammate to your microbiome by eating healthy, drinking lots of water, and getting regular exercise to keep in tip-top shape.

A healthy microbiome is like having a superhero team inside of you, always working to keep you healthy and strong!
(even if they just like the food you eat)

Your tiny friends are always there . . .

The Secret Life of Your Microbiome
GLOSSARY
Advanced scientific definitions!

Term	Explanation
Nutrients	Parts of foods that the bod needs to grow, repair itself, and stay healthy!
Microbes	Microbes are tiny living things so small you need a microscope to see.
Bacteria	Bacteria are microbial creatures that can either help or hurt your body, like those in your stomach that help digest, or those that cause infections or grow where they shouldn't.
Viruses	Viruss are tiny germs that invade your body's cells and use the to make more viruses, this might make you sick.
Protists	Protists are single-celled, and can live in water, soil, or inside other living things.
Lactose	Lactose is a natural sugar found in milk.
Serotonin	Serotonin is a chemical in your brain and body that helps control your mood, making you feel calm and happy.

Term	Explanation
Messenger Molecules	Chemicals that carry different signals, (or messages), between different parts of your body to help it work properly.
Immune System	The body's defense team, protecting you from harmful germs and viruses. The soldiers are the macrophages, they devour invaders.
Digestive System	Group of organs like the stomach, gallbladder, digestive tract, intestines, etc. that break down the food you eat into a power source your body can use.
Lactobillus	A type of good bacteria that helps digest food and fight off harmful germs in the stomach.
Bifidobacteria	Helpful bacteria in the intestines that help keep digestion working smoothly.
Intestines	Long hollow tubes in your belly that contain cilia, which help absorb nutrients from food and remove waste.
Stomach Lining	The stomach lining is the protective lining inside your stomach that stops it from being hurt by the strong acid inside.
Antigens	Antigens are things, like germs or pollen, that the body recognizes as foreign and tries to get rid of. This is part of the reason some people get allergies or allergic reactions.

Term	Explanation
UV Rays	UV rays are invisible waves of light from the sun that can harm your skin and eyes if you're not protected.
S. Epidermis	A type of bacteria that lives on your skin and helps protect your skin barrier from bad germs.
Streptococcus Salivarius	A kind of bacteria that lives in your mouth that helps keep your teeth and gums healthy.
Processed Foods	Foods that have been changed in a factory, usually by adding extra ingridients or removing some of their natural parts. Sometimes these aren't that good for you.
Antibiotics	Medicines that kill harmful bacteria in your body when you have an infection. We need to be careful how much we use though, it might kill the good bacteria too!
Probiotics	Probiotics are good bacteria you can eat, like those in yogurt, to help replenish the good bacteria in your body.

www.ingramcontent.com/pod-product-compliance
Lightning Source LLC
Chambersburg PA
CBHW040451220526
45473CB00004B/1597